Birds of Pr

FALCONS

LYDIA LUKIDIS

BLACK
RABBIT
BOOKS

Bolt is published by Black Rabbit Books
P.O. Box 227, Mankato, Minnesota, 56002
www.blackrabbitbooks.com

Marysa Storm, editor; Michael Sellner, designer and photo researcher

Library of Congress Cataloging-in-Publication Data
Names: Lukidis, Lydia, author.
Title: Falcons / by Lydia Lukidis.
Description: Mankato, Minnesota : Black Rabbit Books, [2023] | Series: Bolt. birds of prey | Includes bibliographical references and index. | Audience: Ages 8-12 | Audience: Grades 4-6 | Summary: "Equipped with strong senses and powerful features, falcons rule the skies. They also capture the imagination of readers everywhere. With an engaging design focused around expertly leveled text and colorful infographics that support visual literacy, introduce reluctant and beginning readers to the life cycle, features, and diets of these powerful birds of prey" Provided by publisher.
Identifiers: LCCN 2020017987 (print) | LCCN 2020017988 (ebook) | ISBN 9781623105532 (hardcover) | ISBN 9781644664841 (paperback) | ISBN 9781623105594 (ebook)
Subjects: LCSH: Falcons–Juvenile literature.
Classification: LCC QL696.F34 L84 2023 (print) | LCC QL696.F34 (ebook) | DDC 598.9/6-dc23
LC record available at https://lccn.loc.gov/2020017987
LC ebook record available at https://lccn.loc.gov/2020017988

Image Credits

123RF: photoncatcher, 17; Alamy: Allexxandar, 3, 12; Arterra Picture Library, 11; Harry Collins, 20; imageBROKER, 11, 21; Tom Meaker, 32; CARRI: SCOTT RASHID/ CARRI, 21; Dreamstime: Makoto Hasegawa, 4–5; Vasyl Helevachuk, 8–9; Getty: Education Images, 14–15; Jeff Wendorff, 28–29; iStock: jacquesvandinteren, 16; Minden Pictures: Murray Cooper, 21; Shutterstock: Aksenova Natalya, 25; BearFotos, 25; BIOSPHOTO, 26; Christopher Chambers, 11; CRS PHOTO, 6; Dennis Jacobsen, 26–27; Faan, Cover; Gallinago_media, 7; Harry Collins Photography, 11; JRP Studio, 21; Koshevnyk, 7; Lia Li, 6; L-N, 22–23; Melinda Fawver, 21; panda3800, 25; SelimBT, 16; specnaz, 25; Sriram Bird Photographer, 12; S.Zykov, 7; Weranut, 1, 18–19, 31; WildlifeWorld, 21; withGod, 25

Contents

CHAPTER 1

A CLOSER Look at Falcons

A peregrine falcon glides through the air. It's on the hunt. The bird spots a pigeon flying below. It follows the pigeon for a while. Then the falcon swoops down at lightning speed. Using its sharp beak, the falcon kills its **prey**. The falcon then brings its meal to a tree branch. Time for dinner!

COMPARING WINGSPANS

AMERICAN KESTREL

20 to 24 INCHES
(51 to 61 centimeters)

Strong Birds

Falcons are powerful birds of prey. They eat only meat, and they're excellent hunters. There are more than 60 different kinds of falcons. They all have long, thin, pointed wings. Their beaks are short and curved. Falcons use them to kill prey. Each falcon foot has four sharp claws called talons. The birds use them to carry prey.

PARTS OF A FALCON

POINTED WINGS
SHORT TAIL
TALONS

Different Types

Scientists split falcons into four groups. Kestrels are small falcons. Their feathers are brown with dark spots. Hobbies are a bit bigger with many dark gray or black feathers. Peregrines are very big and more powerful. Most of them have black feathers on their heads. Hierofalcons are also large falcons. They have brown or red feathers.

KESTRELS

HOBBIES

PEREGRINES

HIEROFALCONS

Peregrine falcons are the fastest falcons. They can reach speeds of more than 200 miles (322 kilometers) per hour.

Falcon Superpowers

All falcons have amazing eyesight. Scientists believe the birds can see eight times better than humans. They can spot little animals from more than 1 mile (2 km) away.

Falcons' incredible speed is another superpower. In fact, falcons are the fastest birds in the world. Their **aerodynamic** bodies help them zip through the air. Their long, pointed wings are strong. They help falcons fly quickly.

CHAPTER 2

At HOME and on the Hunt

Falcons spend their days hunting. They often go after animals in the sky, such as smaller birds. Falcons also eat insects, small **mammals**, and **reptiles**.

Not all falcons hunt the same way. Often, peregrines will dive after prey in the sky. They strike quickly. Other falcons, such as kestrels, will hover over fields. They wait for the perfect moment to pounce on insects.

Falcon Habitats

DESERTS

FORESTS

MOUNTAINS

Making Homes

Falcons live all around the world. They fly in a wide range of habitats. Some make deserts, forests, or mountains their homes. Others live in the Arctic or in cities.

Falcon nests are usually simple. The birds often find a ledge on a cliff. Then they scrape the earth to create a little dent for their eggs. Sometimes they steal nests from other birds.

Falcons live on every continent except Antarctica.

ASIA
EUROPE
AFRICA
AUSTRALIA
ANTARCTICA

CHAPTER 3

FAMILY Life

Most types of falcons prefer flying alone. But they do come together to have babies. Many falcons choose one **mate**. They meet up every year. In spring, the females lay between one and seven eggs. This group of eggs is called a clutch. About a month later, the eggs hatch.

Comparing Clutch Sizes

aplomado falcon
1–3 eggs

peregrine falcon
2–6 eggs

gyrfalcon
2–7 eggs

American kestrel
2–7 eggs

Baby Falcons

Baby falcons are helpless. Their parents feed and take care of them. The parents must protect the babies from **predators**, such as foxes. At about one month old, the chicks begin to practice flying. A few months later, they can fly on their own.

CHAPTER 4

PREDATORS and Other Dangers

Falcons are **fierce**. Not many animals hunt them. Their large size keeps attackers away. The birds also protect themselves with their talons and beaks. However, bears, wolves, and some larger birds eat falcons.

Falcon Food Chain

This food chain shows what eats falcons.
It also shows what falcons eat.

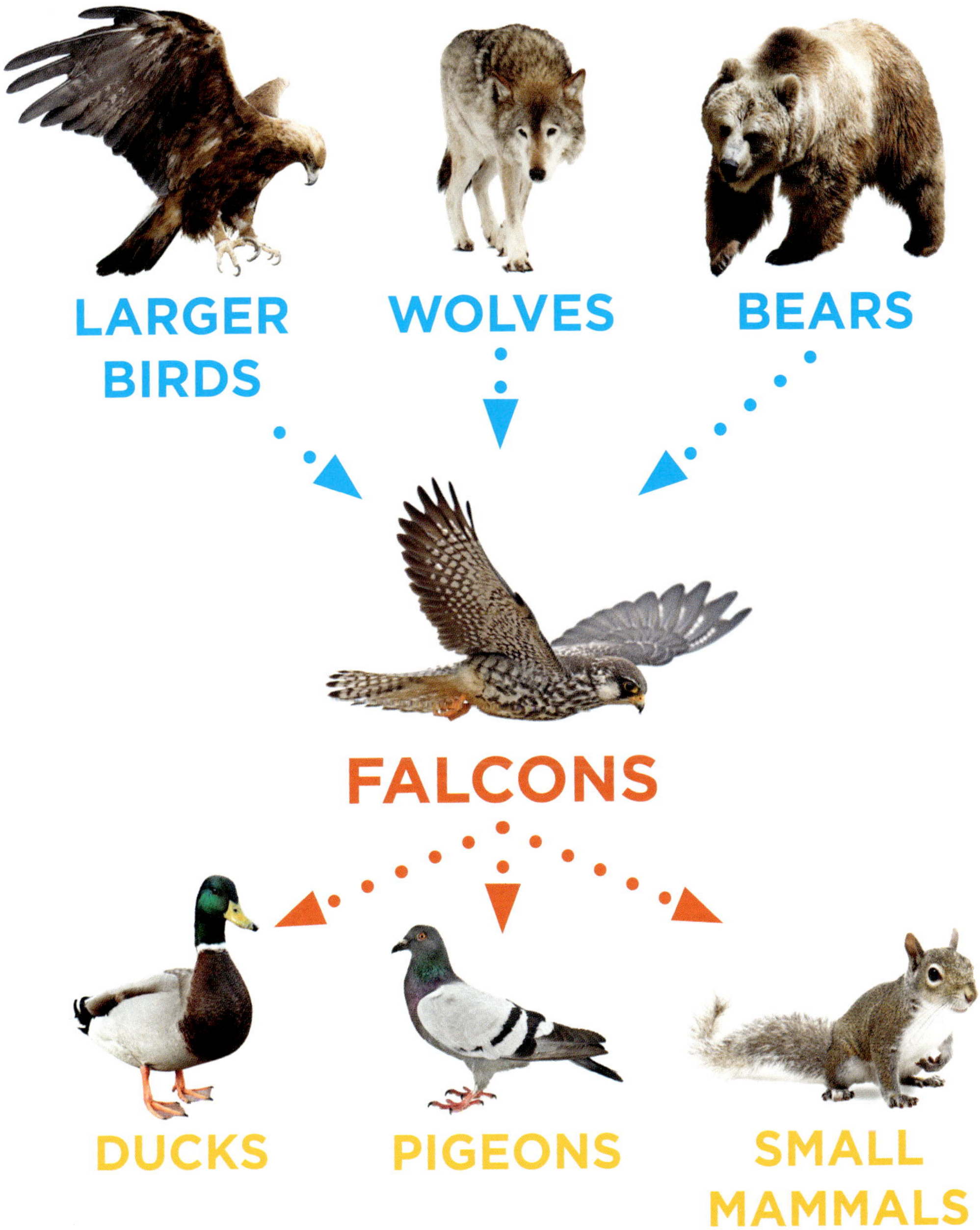

Years ago, people created a poison to keep bugs from crops. But falcons ate the bugs. The poisoned bugs killed them. Falcons began to disappear from North America. The United States **banned** the poison.

Human Threats

Humans are also a threat to falcons. They destroy falcon habitats to build homes. The birds have fewer places to nest and hunt.

Every single animal has its place in the food chain. Falcons eat many animals. They're food for other animals too. Falcons are needed to keep nature in balance.

By the NUMBERS

4
times per second

HOW FAST A PEREGRINE FALCON CAN BEAT ITS WINGS

1/4
OF ITS WEIGHT

HOW MUCH A FALCON CAN EAT AT ONE TIME

3 to 6
OUNCES
(85 to 170 GRAMS)

how much an American kestrel weighs

13 YEARS

AVERAGE GYRFALCON LIFE SPAN

GLOSSARY

aerodynamic (air-oh-dahy-NAM-ik)—something that is shaped in a way that helps it move easily through air

ban (BAHN)—to forbid by law

fierce (FEERS)—very strong or powerful

mammal (MAH-muhl)—a type of animal that feeds milk to its young and usually has hair or fur

mate (MAYT)—one of a pair that joins together to produce young

predator (PRED-uh-tuhr)—an animal that eats other animals

prey (PRAY)—an animal hunted or killed for food

reptile (REP-tahyl)—a cold-blooded animal that breathes air and has a backbone; most reptiles lay eggs and have scaly skin.

BOOKS

Gleisner, Jenna Lee. *Falcons.* Raptor World. Minneapolis: Jump!, Inc., 2020.

Riggs, Kate. *Falcons.* Amazing Animals. Mankato, MN: The Creative Company, 2023.

Walker, Alan. *Birds of Prey.* Built to Survive. New York: Crabtree Publishing, 2022.

WEBSITES

Falcon
kids.britannica.com/kids/article/falcon/353117

Peregrine Falcon
kids.nationalgeographic.com/animals/birds/facts/peregrine-falcon/

Ten Fun Facts about Falcons, the Birds
www.smithsonianmag.com/science-nature/10-fun-facts-about-falcons-180962042/

INDEX